Programming Arduino with LabVIEW

Build interactive and fun learning projects with Arduino using LabVIEW

Marco Schwartz

Oliver Manickum

PUBLISHING

BIRMINGHAM - MUMBAI

Programming Arduino with LabVIEW

First published: January 2015

Production reference: 1210115

Published by Packt Publishing Ltd.
Livery Place
35 Livery Street
Birmingham B3 2PB, UK.

ISBN 978-1-84969-822-1

www.packtpub.com

Credits

About the Authors

Marco Schwartz is an electrical engineer, entrepreneur, and blogger. He has a master's degree in electrical engineering and computer science from SUPELEC in France and a master's degree in micro engineering from the EPFL in Switzerland.

He has more than 5 years of experience working in the domain of electrical engineering. His interests gravitate around electronics, home automation, the Arduino and Raspberry Pi platforms, open source hardware projects, and 3D printing.

He also runs several websites on Arduino, including the `http://www.openhomeautomation.net/` website, which is dedicated to building home automation systems using open source hardware.

He has written another book called *Arduino Home Automation Projects*, *Packt Publishing*, on home automation and Arduino and also published a book called *Internet of Things with the Arduino*, on how to build Internet-of-Things projects with Arduino.

Oliver Manickum has been working in the embedded development scene for almost 20 years. His favorite development platform is Arduino. He has delivered thousands of projects and is a big fan of ATMEL and the Arduino platform. He currently writes high-performance games on mobile platforms; however, developing prototypes with Arduino is his main hobby.

He has also reviewed *Netduino Home Automation Projects, Matt Cavanagh*.

I would like to thank my wife, Nazia Osman, for her patience while I was building devices that would sometimes burn down parts of our house, over and over again.

About the Reviewers

Adith Jagadish Boloor is an undergraduate student at the School of Mechanical Engineering at Purdue University, West Lafayette. He was born and brought up in the beautiful coastal city of Mangalore, India. Having lived there for 18 years, he came to the United States of America to pursue his higher education, with the desire to acquire new skills pertaining to the latest technological developments, and with this knowledge, he hopes to revolutionize the robotics sector.

Having built a couple of robots in his high-school days, his primary interest lies in the field of robotics. However, he occasionally occupies himself in areas that are still at their infancy, such as 3D Printing and Speech Recognition. More recently, he has begun his exploration in home automation, wireless networking, the Internet of Things, and smart security systems.

His passion for kindling the benefits of technology is what drives him towards open source and to create a smarter planet.

Aaron Srivastava is a biomedical engineer from North Carolina State University. He is currently working on a neurosurgery project to aid patients undergoing spinal cord stimulation treatments. His main interests are in entrepreneurship, business development, and programming languages. Aaron also does web designing, on the side, as a hobby.

Fangzhou Xia is a dual-degree senior student at University of Michigan, with a background in both mechanical engineering and electrical engineering. His areas of interest in mechanical engineering are system control, product design, and manufacturing automation. His areas of interest in electrical engineering are web application development, embedded system implementation, and data acquisition system setup.

www.PacktPub.com

Support files, eBooks, discount offers, and more

For support files and downloads related to your book, please visit www.PacktPub.com.

Did you know that Packt offers eBook versions of every book published, with PDF and ePub files available? You can upgrade to the eBook version at www.PacktPub.com and as a print book customer, you are entitled to a discount on the eBook copy. Get in touch with us at service@packtpub.com for more details.

At www.PacktPub.com, you can also read a collection of free technical articles, sign up for a range of free newsletters and receive exclusive discounts and offers on Packt books and eBooks.

https://www2.packtpub.com/books/subscription/packtlib

Do you need instant solutions to your IT questions? PacktLib is Packt's online digital book library. Here, you can search, access, and read Packt's entire library of books.

Why subscribe?

- Fully searchable across every book published by Packt
- Copy and paste, print, and bookmark content
- On demand and accessible via a web browser

Free access for Packt account holders

If you have an account with Packt at www.PacktPub.com, you can use this to access PacktLib today and view nine entirely free books. Simply use your login credentials for immediate access.

Table of Contents

Preface

Arduino is a powerful electronics prototyping platform used by millions of people around the world to build amazing projects. Using Arduino, it is possible to easily connect sensors and physical objects to a microcontroller, without being an expert in electronics.

However, using Arduino still requires us to know how to write code in C/C++, which is not easy for everyone. This is where LabVIEW comes into play. LabVIEW is software used by many professionals and universities around the world, mainly to automate measurements without having to write a single line of code.

Thanks to a module called LINX, it is actually very easy to interface Arduino and LabVIEW. This means that we will be able to control Arduino projects without having to type a single line of code. The possibilities are endless, and in this book, we will focus on several exciting projects in order for you to discover the key features of the LabVIEW Arduino interface.

What this book covers

Chapter 1, *Welcome to LabVIEW and Arduino*, introduces you to the Arduino platform and the LabVIEW software.

Chapter 2, *Getting Started with the LabVIEW Interface for Arduino*, shows you how to install and use the LabVIEW interface for Arduino via the LINX module.

Chapter 3, *Controlling a Motor from LabVIEW*, explains how to make your first real project with Arduino and LabVIEW by controlling a DC motor from LabVIEW.

Chapter 4, *A Simple Weather Station with Arduino and LabVIEW*, talks about how to automate measurements from several sensors that are connected to the Arduino platform.

Chapter 5, Making an XBee Smart Power Switch, shows you how to make our own *do-it-yourself* (DIY) version of a smart wireless power switch. We will make a device that can control electrical devices, measure their current consumption, and control the whole power switch from LabVIEW.

Chapter 6, A Wireless Alarm System with LabVIEW, helps you connect motion sensors to an Arduino board and monitor their state remotely via LabVIEW to create a simple alarm system.

Chapter 7, A Remotely Controlled Mobile Robot, teaches you how to use everything you learned so far to control a small mobile robot from LabVIEW. You will be able to wirelessly move the robot and also continuously measure the distance in front of the robot.

What you need for this book

For this book, you will mainly need the LabVIEW software that is available for all major operating systems. You can either buy it or download an evaluation version for free.

You will also need the LINX module to interface LabVIEW and Arduino, which we will see how to set up and use in *Chapter 2, Getting Started with the LabVIEW Interface for Arduino* of the book.

Who this book is for

This book is for people who already have some experience with the LabVIEW software and who want to use the Arduino platform. For example, if you want to automate measurements from sensors and control physical objects with Arduino, but without writing Arduino code, this book is for you.

It is also for people who already have some knowledge of the Arduino platform and who want to learn another way to control their Arduino projects, using LabVIEW instead of coding.

Conventions

In this book, you will find a number of styles of text that distinguish between different kinds of information. Here are some examples of these styles, and an explanation of their meaning.

New terms and **important words** are shown in bold. Words that you see on the screen, in menus or dialog boxes for example, appear in the text like this: "clicking the **Next** button moves you to the next screen."

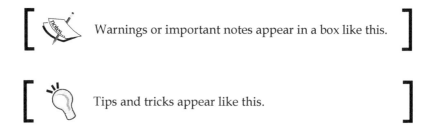

Warnings or important notes appear in a box like this.

Tips and tricks appear like this.

Reader feedback

Feedback from our readers is always welcome. Let us know what you think about this book—what you liked or may have disliked. Reader feedback is important for us to develop titles that you really get the most out of.

To send us general feedback, simply send an e-mail to feedback@packtpub.com, and mention the book title via the subject of your message.

If there is a topic that you have expertise in and you are interested in either writing or contributing to a book, see our author guide on www.packtpub.com/authors.

Customer support

Now that you are the proud owner of a Packt book, we have a number of things to help you to get the most from your purchase.

Downloading the example code

You can download the example code files for all Packt books you have purchased from your account at http://www.packtpub.com. If you purchased this book elsewhere, you can visit http://www.packtpub.com/support and register to have the files e-mailed directly to you.

Downloading the color images of this book

We also provide you a PDF file that has color images of the screenshots/diagrams used in this book. The color images will help you better understand the changes in the output. You can download this file from: `http://www.packtpub.com/sites/default/files/downloads/8221OT_ColorImages.pdf`.

Errata

Although we have taken every care to ensure the accuracy of our content, mistakes do happen. If you find a mistake in one of our books—maybe a mistake in the text or the code—we would be grateful if you would report this to us. By doing so, you can save other readers from frustration and help us improve subsequent versions of this book. If you find any errata, please report them by visiting `http://www.packtpub.com/submit-errata`, selecting your book, clicking on the **errata submission form** link, and entering the details of your errata. Once your errata are verified, your submission will be accepted and the errata will be uploaded on our website, or added to any list of existing errata, under the Errata section of that title. Any existing errata can be viewed by selecting your title from `http://www.packtpub.com/support`.

Piracy

Piracy of copyright material on the Internet is an ongoing problem across all media. At Packt, we take the protection of our copyright and licenses very seriously. If you come across any illegal copies of our works, in any form, on the Internet, please provide us with the location address or website name immediately so that we can pursue a remedy.

Please contact us at `copyright@packtpub.com` with a link to the suspected pirated material.

We appreciate your help in protecting our authors, and our ability to bring you valuable content.

Questions

You can contact us at `questions@packtpub.com` if you are having a problem with any aspect of the book, and we will do our best to address it.

1
Welcome to LabVIEW and Arduino

National Instruments Corporation, NI, is a world leader when it comes to automated test equipment and virtual instrumentation software. LabVIEW is a product that they have developed, and it is being used in many labs throughout the world. LabVIEW, which stands for Laboratory Virtual Instrument Engineering Workbench, is programmed with a graphical language known as G; this is a dataflow programming language. LabVIEW is supported by **Visual Package Manager (VIPM)**. VIPM contains all the tools and kits to enhance the LabVIEW product.

Arduino is a single-board microcontroller. The hardware consists of an open source hardware board that is designed around the Atmel AVR Microcontroller. The intention of Arduino was to make the application of interactive components or environments more accessible. Arduinos are programmed via an **integrated development environment (IDE)** and run on any platform that supports Java. An Arduino program is written in either C or C++ and is programmed using its own IDE.

Welcome to programming Arduino with LabVIEW. During the course of this book, we will take you through working with Arduino through NI's LabVIEW product. The following are what you will need:

- A Windows or Mac-based machine
- Arduino (Uno preferred)
- LabVIEW 13 for students (or any other LabVIEW 13 distribution)

We will work with Servos, LEDs, and Potentiometers in both analog and digital configurations.

What makes Arduino ideal for LabVIEW

The Arduino community is extremely vast with thousands and even hundreds of thousands of projects that can be found using simple searches on Google. Integrating LabVIEW with Arduino makes prototyping even simpler using the GUI environment of LabVIEW with the Arduino platform.

Officially, LabVIEW will work with the Uno and Mega 2560; however, you should be able to run it on other Arduino platforms such as the Nano. Building your own Uno board is just as simple as linking up the Arduino to LabVIEW. For detailed instructions on how to build your own Arduino Uno, check out the following URL: `http://www.instructables.com/id/Build-Your-Own-Arduino/`.

Significance of using LabVIEW

LabVIEW is a graphical programming language built for engineers and scientists. With over 20 years of development behind it, it is a mature development tool that makes automation a pleasure.

The graphical system design takes out the complexity of learning C or C++, which is the native language of Arduino, and lets the user focus on getting the prototype complete.

LabVIEW significantly reduces the learning curve of development, because graphical representations are more intuitive design notations than text-based code. Tools can be accessed easily through interactive palettes, dialogs, menus, and many function blocks known as **virtual instruments (VIs)**. You can drag-and-drop these VIs onto the **Block Diagram** to define the behavior of your application. This point-and-click approach shortens the time it takes to get from the initial setup to a final solution.

Skills required to use LabVIEW and Arduino

With LabVIEW primarily being designed for and targeted at scientists and engineers, it has not excluded itself from being used by hobbyists. Users who have zero programming skills have been able to take entire projects to completion by just following the intuitive process of dragging controls onto the diagram and setting it up to automate.

We have designed this book to be completely intuitive, using parts that can be easily found at your local electronic store.

To get additional support when using LabVIEW with Arduino, have a look at their forum at `https://decibel.ni.com/content`.

Downloading LabVIEW

To download or purchase LabVIEW, head out to `http://www.ni.com/trylabview/`. LabVIEW can also be purchased with an Arduino Uno bundle from SparkFun. At the time of writing this book, the URL for this bundle is `https://www.sparkfun.com/products/11225`.

 If you did not download LabVIEW, do so now. To try LabVIEW without purchasing it, click on **Launch LabVIEW**.

To install the product, click on all the default options. Note that the Arduino plugin is not found in the initial install of LabVIEW.

Once LabVIEW is installed, launch the Visual Package Manager.

The VIPM will now launch. The VIPM application will look like this:

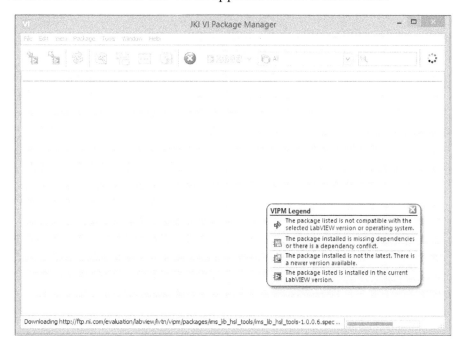

The VIPM will start downloading references to the package bundles into its repository. The status bar is located at the bottom of the application; when the references are downloaded, the status bar will switch to **Ready**.

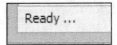

Downloading the Arduino IDE

To download the Arduino IDE, go to `http://arduino.cc/en/main/software`. This book covers the Windows versions of LabVIEW and Arduino; however, the Mac versions will work just as well.

Click on **Windows Installer** to download the Windows version of the Arduino IDE.

 At the time of writing this book, the current version of Arduino IDE is 1.5.8.

To install the product, click on all the default options.

Once the Arduino IDE is installed, click on the shortcut shown here to launch the application:

The Arduino IDE will launch with the following screen:

Now that the default settings for each of the applications are set up and launched, we are ready to start programming in each application.

Summary

In this chapter, you learned more about LabVIEW and Arduino. We also installed all the software that we need to get LabVIEW and the Arduino IDE up and running. In the next chapter, we will get the Arduino package for LabVIEW installed and upload a basic sketch to the Arduino board.

2

Getting Started with the LabVIEW Interface for Arduino

In this second chapter of the book, we will see how to hook up LabVIEW and Arduino. We will connect an Arduino board to our computer, install a special package for LabVIEW, and then control the Arduino board directly from LabVIEW. As an example, we will simply light up the on-board LED of the Arduino Uno board from the LabVIEW interface.

This chapter will really be the foundation for all the projects found in this book, so make sure you follow all the instructions carefully.

Hardware and software requirements

On the hardware side, you will not need a lot for this first project of the book. The only thing you will need is an Arduino Uno board (`https://www.adafruit.com/products/50`). This is the same board that we will use in the rest of the book. You can use other boards as well, such as the Arduino Due or the Arduino Pro. However, I recommend that you stick with the Uno board for the whole book.

On the software side, you will need LabVIEW installed on your computer. For this book, I used LabVIEW 2014 for Windows. Of course, you can use LabVIEW on other platforms such as OS X or Linux. You can also use older versions, as the Arduino package that we will use is compatible with LabVIEW 2011 and above. If you don't have LabVIEW yet, you can find all the information at the following link:

`http://www.ni.com/labview/`

After that, you will need the VIPM. This is free software that interfaces nicely with LabVIEW and allows you to automatically install new packages for LabVIEW.

You can download it from the following link:

`http://jki.net/vipm/download`

If you encounter an error during the installation that says a version of the software is already installed, make sure that you uninstall the old version first and then retry.

Finally, you will need to install the LINX package, which is a new package replacing the old **LabVIEW Interface for Arduino (LIFA)**.

You can get it at the following URL:

`http://sine.ni.com/nips/cds/view/p/lang/en/nid/212478`

On this page, you will find a link to download the package.

LINX - LVH

Interface With Common Embedded Platforms

E-mail this Page | Print | PDF | Rich Text

[+] Enlarge Picture

- Interface with chipKIT, Arduino, and other embedded platforms
- Access peripherals such as DIO, AIO, PWM, SPI, and I2C from LabVIEW
- Take advantage of support for many common sensors
- Communicate over USB, serial, Ethernet, and wireless
- Easily deploy LINX code to NI myRIO and access I/O
- Quickly add a GUI to an embedded project

Download

Follow this link, and you will be taken to another page with the direct link for the VI package manager. Click on the **Download Toolkit** button to start the installation process:

The VI package manager should open automatically and install the LINX package.

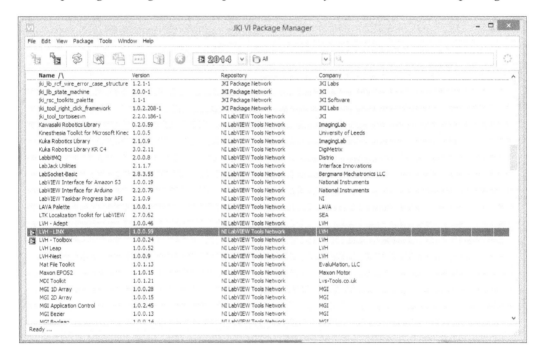

If this does not work and you get an error, it may be linked to the download servers, which may have an issue. In this case, simply retry the procedure, and it should work.

Setting up LabVIEW and LINX

We will now set up LabVIEW and the LINX package so that all the projects of this book can work correctly. Perform the following steps:

1. First, start LabVIEW. Don't create any project, but click on **Tools** and then on **Options**.

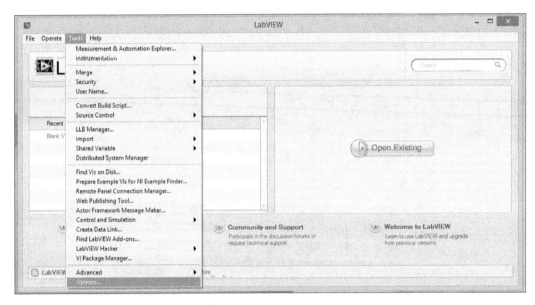

2. You will be taken to the **Options** window of LabVIEW, where you can set all your preferences. Right now, we have to go to the **VI server** menu.

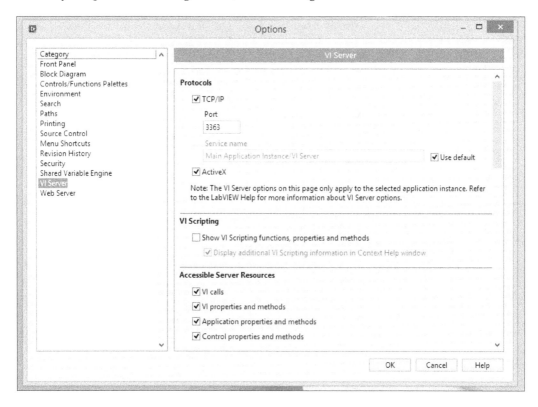

3. You can see that there are some options that you can change here. Change all the options so that they match the options shown in the preceding screenshot.

4. After that, we have to do the same on the VI Package Manager so that both LabVIEW and the Package Manager can talk to each other. On systems like Windows, it was automatically done, but it was not the case on OS X, for example. To do so, simply open the Package Manager, go to the **Tools | Options** menu, and then click on the **LabVIEW** icon.

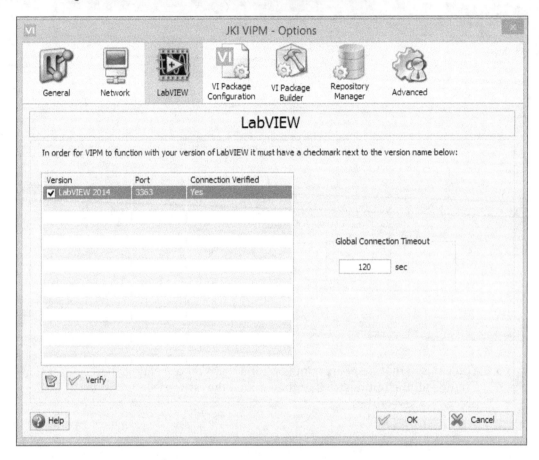

5. In this menu, make sure that the **Port** value next to your LabVIEW installation is the same as the one you defined inside LabVIEW. Correct it here if it is not the case, and confirm.

Testing the installation

We are now ready to test our LabVIEW/LINX installation and start testing our LabVIEW interface for Arduino.

The first thing that you need to do is go to the main LabVIEW window; then, click on **Tools** and then on **LabVIEW Hacker**, which is the link to access the LINX interface. Then, click on **LINX**, and finally, click on **LINX Firmware Wizard**.

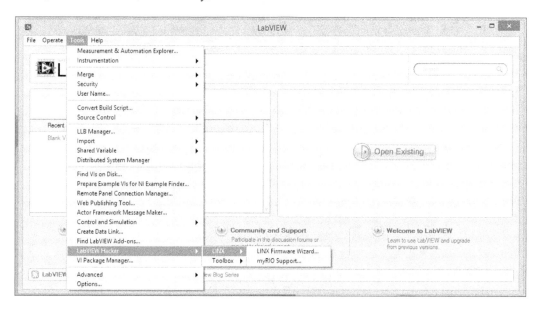

This will take you to the LINX graphical interface that we will use to configure our Arduino board for the project. Note that this step has to be done only one time; once the right software is loaded into the Arduino board, you won't have to touch it again.

The wizard starts by asking us which board we are going to use. Configure this first page by selecting the same settings as shown in the following screenshot:

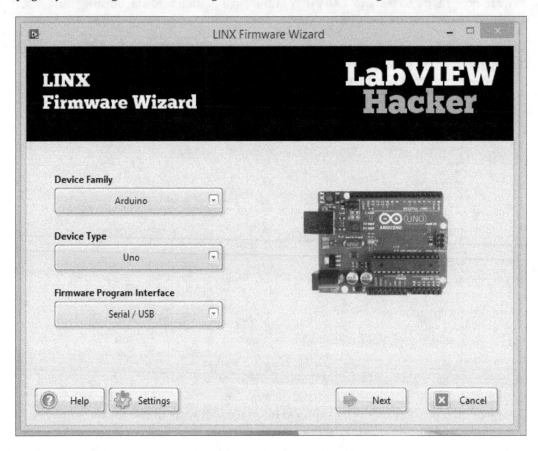

After that, you will be prompted to select the Serial Port on which you want the interface to communicate. As I only had one Arduino board connected at that time, I could only select the port that Windows calls COM4. Of course, this will entirely depend on your operating system.

A very simple way to find the COM or Serial Port that corresponds to your Arduino board is to look at the list of proposed Serial ports. Then, disconnect your board and see which Serial Port disappeared; this is the one that corresponds to your board.

Finally, confirm your choice of Serial Port, and start uploading the firmware on the Arduino board.

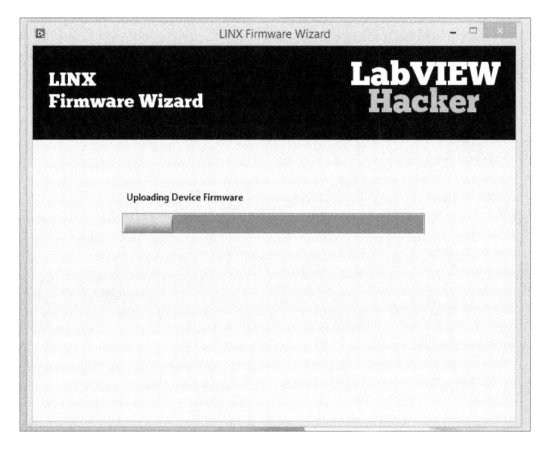

Congratulations! You are now ready to use the LINX interface to control your Arduino board.

If you had an issue at this step, you might have to install the NI-VISA package, which you can download from this link:

`http://www.ni.com/download/ni-visa-4.3/988/en/`

At the end of this setup, LINX will offer to open an example program. Accept this offer, and you will be taken to a new VI.

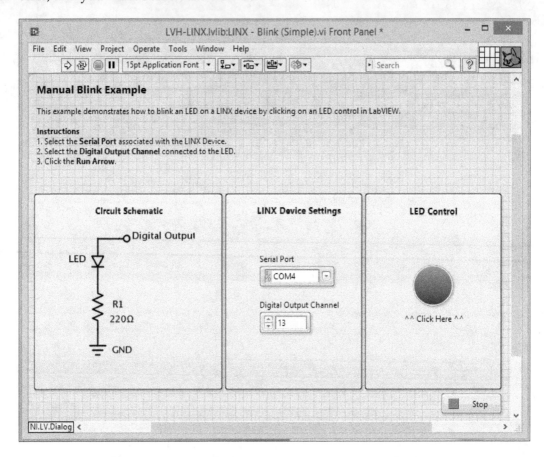

This is called the **Front Panel** of this example project from which you can control the project. As you can see, this VI is really simple, as you can just control the value of a digital pin of the Arduino by clicking on the green button on the right-hand side.

There are two things you need to modify here before you can start the VI. First, you need to set the correct Serial Port in the **Serial Port** box. Just start by typing the name of your port, and it will autocomplete what you are writing.

Then, you need to set which pin you want to control. I simply used pin number **13** here, as it is already connected to the on-board LED on the Arduino Uno board. If you choose any other pin, you will be able to build a simple circuit on your board, as shown in the illustration on the left-hand side of the preceding screenshot.

Let's now use the VI. To do so, simply click on the small arrow on the toolbar. Then, wait for a while. Indeed, the VI will now try to initialize the communication with the Arduino board. If you click on something immediately, it can produce an error. You will know that the initialization process is complete when the Arduino board Serial Port LEDs (TX & RX) are both turned on. Then, click on the green button; you will see that the on-board LED on the Arduino board is immediately turning on or off.

Let's go a bit further and see what is behind that sketch. The details are beyond the scope of this chapter, but it can be interesting to see what is going on at this stage. To do so, go to **Window** and then click on **Show Block Diagram**. Note that you can also use the *Ctrl + E* shortcut to switch between **Front Panel** and **Block Diagram**. This will open the following window:

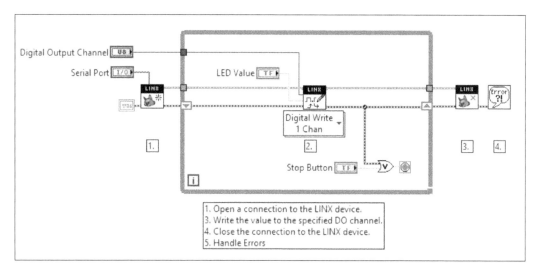

This is the **Block Diagram** window for this project, which is basically what is going on behind the scenes. Some of the components are linked to elements of **Front Panel**, such as the **Serial Port** value. You can see that the core of the project is this **Digital Write** module that we use to send commands to the Arduino board.

For now, we really just wanted to have an overview of what is done in this diagram. In the following chapters of the book, you will see how to build such block diagrams from scratch to build your own projects.

Summary

Let's summarize what we saw in this chapter. You learned how to install the software components that are required for the whole book, such as the VI package manager and the LINX interface for Arduino. This way, you will be able to control Arduino boards from LabVIEW.

We also saw a basic example of a VI used to control an Arduino board, and as an application, we controlled the on-board LED on the Arduino Uno board.

At this stage, it is really important that you perform every step of this chapter correctly, as we will build all the projects in the book based on these steps. If you want to go a little further, you can play with the Block Diagram window of this chapter and modify it a bit. You can also play with the examples that come with the LINX package, which are located in the examples folder of your LabVIEW installation folder.

In *Chapter 3, Controlling a Motor from LabVIEW*, you will use what you have learned so far to create your first useful application using LabVIEW and Arduino.

3

Controlling a Motor from LabVIEW

In this chapter, we will write our first VI (LabVIEW program) from scratch. As an example, we will control a DC motor that is connected to the Arduino board. We will build the VI from scratch and then control the direction and speed directly from the LabVIEW graphical interface.

Hardware and software requirements

On the hardware side, you will first need an Arduino Uno board.

For the motor, I chose a small 5V DC motor from Amazon. You can choose any brand that you want for the motor; the important thing is that it has to be rated to work at 5V so that it can be powered directly from Arduino. You can also get a motor that uses higher voltages or currents, but you will need to modify the hardware configuration slightly.

You will also need the L293D motor driver to control the motor from Arduino. This is a dedicated chip that we will use to easily control the motor from LabVIEW. You can also use an alternative to this chip; for example, you can use an Arduino shield that already integrates similar chips on the board. This is, for example, the case of the official Arduino motor shield, which integrates the L298D chip. However, you would need to modify the code slightly if you are using a shield instead of the chip alone.

Finally, you will need a breadboard and jumper wires to make all the connections.

This is a list of all the components required for this chapter, along with the links to find them on the Web:

- Arduino Uno (https://www.adafruit.com/products/50)
- L293D (https://www.adafruit.com/product/807)
- DC motor (http://www.amazon.com/Motor-5V-80mA-200mA-torque/dp/B001DAYVA6)
- Jumper wires (https://www.adafruit.com/products/1957)
- Breadboard (https://www.adafruit.com/products/64)

On the software side, you will need to have LabVIEW and the LINX package installed. If this is not done yet, refer to *Chapter 2, Getting Started with the LabVIEW Interface for Arduino*, to follow all the required steps.

Hardware configuration

Let's now see how to assemble the different components of the project. This schematic will help you visualize the connections between the different components:

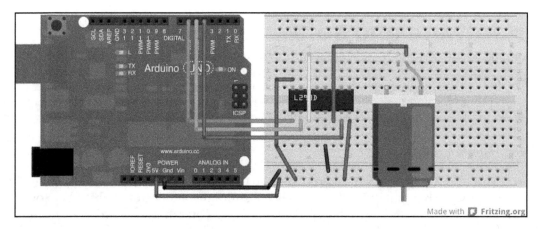

To assemble the components follow the steps:

1. First, put the L293D chip in the middle of the breadboard.
2. Then, take care of the power supply; connect the upper-left pin and the lower-right pin of the L293D chip to the Arduino 5V pin.
3. Then, connect one of the pins at the lower center of the chip to the Arduino GND pin.

4. After that, connect the command signals coming from the Arduino, which will be on pins 4, 5, and 6, and the Arduino Uno board.

5. Finally, connect the DC motor to the L293D chip, as shown in the schematic.

To help you out, here is a link to the pins' configuration of the L293D chip:

`http://users.ece.utexas.edu/~valvano/Datasheets/L293d.pdf`

This is what it should look like at the end:

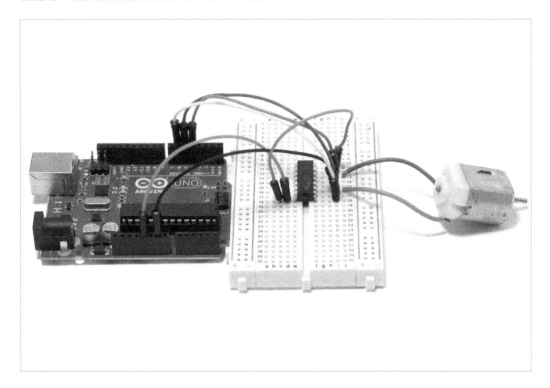

When this is done, you can move to the next step; building the VI in LabVIEW to control the DC motor.

Writing the LabVIEW program

We will now write a new LabVIEW program from scratch so that you can see how the LINX interface for Arduino is working. To start the process, open LabVIEW and create a new blank VI.

We already saw in the previous chapter that there are two main views in LabVIEW: **Front Panel** and **Block Diagram**. In your new blank VI, these two views will be empty. We will first take care of **Block Diagram**, where we will add the elements to control the Arduino board.

Note that we will directly learn about LabVIEW and Arduino by building our first project.

If you want to learn more about the LabVIEW software first, you can visit this link:

`http://www.ni.com/getting-started/labview-basics/`

To learn the basics of Arduino first, the best option is to explore the official Arduino website:

`http://arduino.cc`

The first thing we will place on the blank VI is a While Loop that you can just drag-and-drop from the **Functions** menu (which you can call at any moment with a right-click). The While Loop can be found in the **Structures** submenu. This loop is required for any Arduino board you want to control via LINX, and all the Arduino commands will need to be placed inside this loop.

This is how it will look on your VI:

Downloading the example code

You can download the example code files for all Packt books you have purchased from your account at http://www.packtpub.com. If you purchased this book elsewhere, you can visit http://www.packtpub.com/support and register to have the files e-mailed directly to you.

After that, we will place our first elements from the LINX package. The first elements we need to place are the LINX initialize and stop elements, which are necessary to tell the software where to start and where to stop. You can find both boxes in the functions panel by going to the **LabVIEW Hacker** submenu.

From the same submenu, place two **Digital Write** blocks (which will be used to control the motor direction) and one **PWM** block (which will be used to control the motor speed). Note that you can find these blocks under the **Peripherals** menu. This is the result:

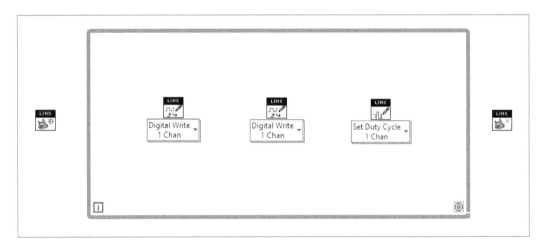

We need a **PWM** block here to control the speed of the motor. PWM stands for Pulse Width Modulation and is used to control the motor's speed or to fade LEDs, for example. On the Arduino board, it is an output of the board that can be set from 0 to 255 on some pins of the Uno board.

To learn more about PWM, you can visit the following link:

http://en.wikipedia.org/wiki/Pulse-width_modulation

Now, we need some way to tell LabVIEW in which order we want the sketch to be executed. This is where the error and LINX resource come into play. Simply start from the initialize block on the left-hand side and find the error pin on the block.

Then, connect the error-out pin of this block to the error-in pin of the first digital block and so on till the end block. After that, do the same with the LINX resource pins. I also added a simple error handler at the end of the VI, just after the stop block. This handler can be found under the **Dialog & User Interface** menu.

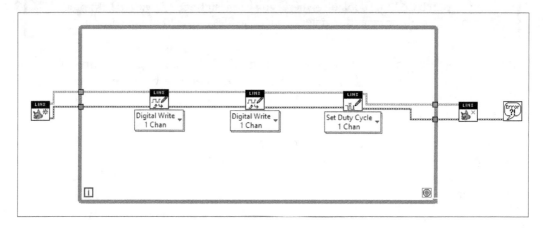

Now that we have the backbone of our project, we will feed the blocks with some inputs. First, add a serial port to the initialize block by going to the serial port pin of the block and right-clicking on it.

Then, go to **Create | Control** to automatically add a serial port input. You will note that the corresponding control is automatically added to **Front Panel** as well. Rename this control to `Serial Port` so that we can identify it in **Front Panel**.

We will also create the same kind of controls for the pins of the blocks we placed earlier. For each block, simply add inputs by right-clicking on the pin's input and then going to **Create | Control**. Also, rename all of these controls so that we know what they mean later in **Front Panel**.

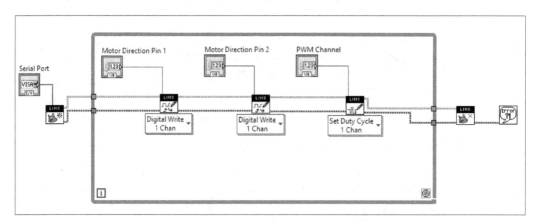

We also need to add an end condition for the While Loop. To do so, we need to connect the little red circle that is located in the bottom-right corner of the While Loop. In this chapter, we will simply connect the error wire directly to this red circle. To do so, just select the input pin of the red circle and connect it to the bottom error wire inside the VI.

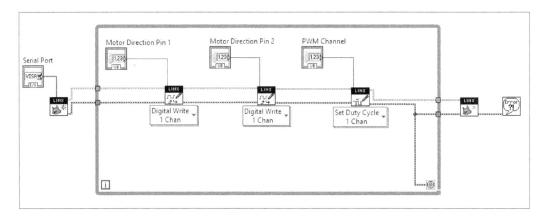

We will now feed the values of the different blocks that we will change from **Front Panel** to control the motor. At this stage, we will keep it simple: we will have some on/off control for the direction and a simple text box for the speed of the motor.

First, let's set the direction that we need to feed on the two first LINX blocks in our VI. The L293D chip requires to be fed with opposite signals on the two direction pins for the motor to rotate in a given direction. For example, when the first **Digital Write** block is on, we want the second one to be off and vice versa.

To do so, we will first create a control block on the first **Digital Write** block, again by right-clicking on the input pin and then going to **Create | Control**. Then, we will go to the Functions menu, in Booleans, choose a Not element, and use it to connect our control to the second **Digital Write** channel. This way, we are sure that these two will always be in opposite states.

Finally, also do the same for the **PWM** block by creating a control for the PWM value. This one will simply be displayed as a text input inside **Front Panel**. We will also rename this pin as `Motor Speed` so that we know what it means in **Front Panel**.

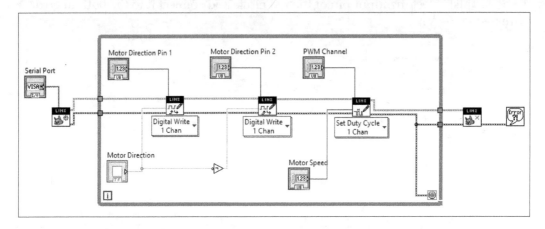

You can now go back to **Front Panel** and have a look at all the elements that were automatically added for you. Organize them a little bit so that it is easier to control the motor.

I simply arranged the **Front Panel** so that all the static controls, such as the serial port and pins, are on the left-hand side (we will modify them only once) and the dynamic controls for the motor are on the right-hand side:

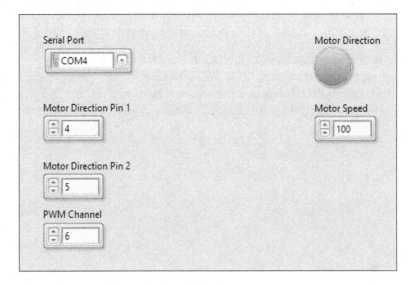

It's now time to test the VI. First, set all the correct pins and your Serial Port, as shown in the preceding image. Then, click on the little arrow in the toolbar to start the VI.

You can now enter a value between 0 and 255 in the Motor Speed input; you will see that the motor starts to rotate immediately. Note that we have to use a value between 0 and 255, as the Arduino Uno PWM output value is coded in 8 bits, so it has 256 values. You can also use the green button to change the direction of the motor.

Upgrading the interface

We now have a basic control for our DC motor, but we can do better. Indeed, it is not so convenient to type in the speed of the motor into **Front Panel** every time you want to modify something. This is why we will introduce another kind of control called a **Knob** control.

To add such a control, start from **Front Panel** and right-click to open the **Controls** panel. Then, go to **Numeric** and select the **Knob** control from the menu.

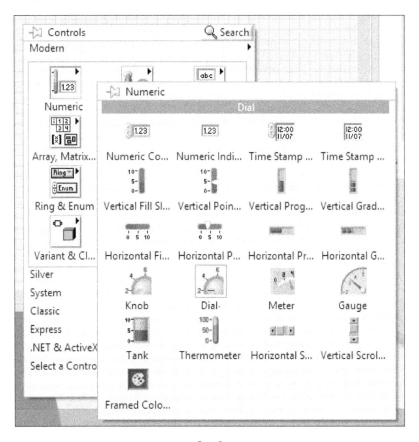

Now, the knob is inserted in **Front Panel**; you can go back to **Block Diagram** where you can remove the old text control from the **PWM** block and connect the new one instead. You can rename it to `Motor Speed` as well.

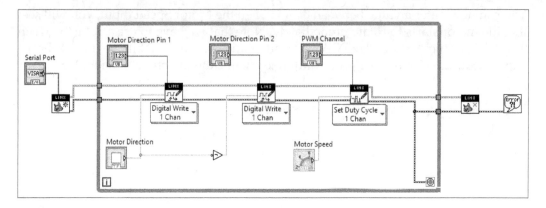

Now, we also need to set the knob so that its output value matches the accepted input of the **PWM** block. Remember that the **PWM** block of LINX accepts values between 0 and 255.

To do so, simply right-click on the **Knob** block and click on **Properties**. In this menu, click on **Scale** and change the minimum and maximum values, as shown in the following screenshot:

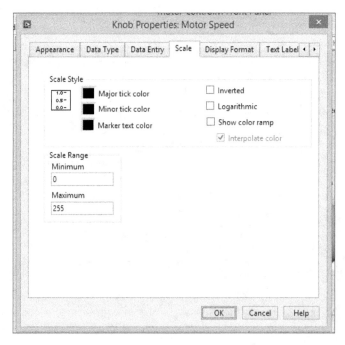

You can now go back to **Front Panel**. You will see that the knob is now displaying the correct values, going from 0 to 255. You can also resize the knob at this point so that it is easier to use.

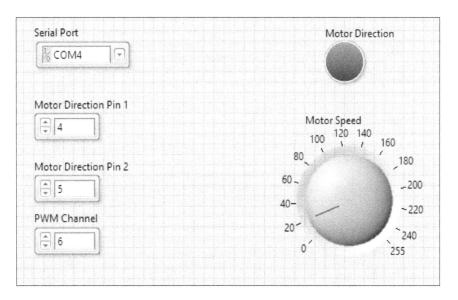

It is now time to test the modified interface. As you did earlier, click on the little arrow inside the toolbar. You can now simply turn the knob to instantly change the rotation speed of the motor.

Summary

Let's summarize what we did in this chapter. We connected a DC motor to Arduino via a dedicated chip to control DC motors. Then, we built an interface in LabVIEW so that we could easily control the direction and speed of this motor. This will be very useful in *Chapter 7, A Remotely Controlled Mobile Robot*, of this book, especially when we will build a robot controlled via LabVIEW.

To go further with what you learned in this chapter, there are some things you can do. You can add more motors to the projects and command them all from a single VI in LabVIEW. You can also use what you learned in this chapter to control simpler components such as LEDs.

This chapter was all about controlling outputs. In the next chapter, we will see how to get data from the inputs of the Arduino board and automate measurements using LabVIEW.

A Simple Weather Station
with Arduino and LabVIEW

In this chapter, we will build a simple weather-measurement station based on Arduino, which will be monitored from LabVIEW.

We will connect a temperature sensor to the Arduino board as well as a light-level sensor. We will connect both of these sensors to the LabVIEW interface so that we can get the measurements from them in real time. Finally, we will use the indicators available in LabVIEW to build a nice graphical interface for the weather-measurement station.

Hardware and software requirements

On the hardware side, you will first need an Arduino Uno board.

We will also use two kinds of sensors for this project: a temperature sensor and light-level sensor. For temperature, we will use a TMP36 sensor, which is an analog temperature sensor that returns a signal depending on the ambient temperature. We will see that there is a block existing in LabVIEW that can automatically convert this output voltage to the ambient temperature.

To measure the ambient light level, we will use a photocell, which is a resistor whose resistance changes with the ambient light. Along with this photocell, we will also need a 10K Ohm resistor. This resistor will be used along with the photocell to convert the ambient light level to an output voltage that will go from 0V to 5V. This voltage will then be converted to a usable variable using the Arduino analog-to-digital converter.

Finally, we will also use a breadboard and some jumper wires to make all the connections between the components.

This is the list of all the components you will need for this project:

- Arduino Uno (`https://www.adafruit.com/products/50`)
- TMP36 (`https://www.adafruit.com/products/165`)
- Photocell and a 10K Ohm resistor (`https://www.adafruit.com/products/161`)
- Jumper wires (`https://www.adafruit.com/products/1957`)
- Breadboard (`https://www.adafruit.com/products/64`)

On the software side, you will need to have LabVIEW and the LINX package installed. If this is not done yet, refer to *Chapter 2, Getting Started with the LabVIEW Interface for Arduino*, to follow all the required steps.

Hardware configuration

Let's now see how to assemble the different components of the project. This schematic will help you visualize the connections between the different components:

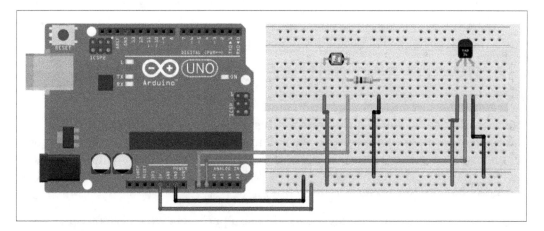

To configure the Arduino Uno follow the steps given:

1. Place the TMP36 sensor and the photocell on the breadboard.
2. Then, connect the power supply from the Arduino board to the breadboard; 5V of the Arduino board goes to the red power rail, and GND goes to the blue power rail.

3. For the TMP36 sensor, there are three pins to connect: VCC, GND, and the output. The output signal is in the middle of the sensor; connect it directly to the analog pin A1 of the Arduino board.

4. Then, looking at the flat part of the sensor, as shown in the schematic, connect the right pin to the blue power rail and the left pin to the red power rail.

5. For the photocell, connect the 10K Ohm resistor in series with the photocell.

6. Then, connect the other pin of the resistor to the blue power rail and the other pin of the photocell to the red power rail of the breadboard.

7. Finally, connect the common pin between the photocell and resistor to the analog pin A0 of the Arduino board.

This is how it should look at the end:

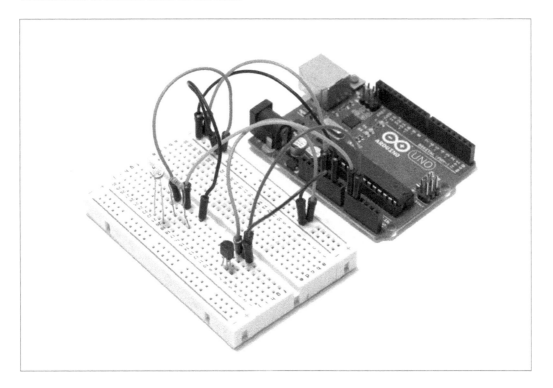

Finally, connect the board to your computer via USB. We are now ready to write the LabVIEW software for our project.

Writing the LabVIEW program

Follow these steps for starting your LabVIEW program:

1. The first thing you need to do here is create a brand new VI in LabVIEW. Then, just as you did in the previous chapter, you need to create a While Loop that will contain all the elements that we will use to interact with the board.

2. As you did in the previous chapter, add an initialize element before the While Loop and a stop element after the While Loop. Also, add a simple error-handling element after the While Loop.

3. We will need two elements inside the While Loop: one to read data from the TMP36 sensor and one to read data from the photocell. Luckily, for us, there is already an element for the TMP36 sensor in the LINX toolbox; this element will automatically calculate the temperature based on the measured data from the sensor.

4. You will find this element in the **Sensors** submenu inside the LINX functions.

5. For the photocell, simply place an **Analog Read** box inside the While Loop. You will also find this box inside the LINX functions; you can access it by navigating to **Peripherals** | **Analog** | **Read**.

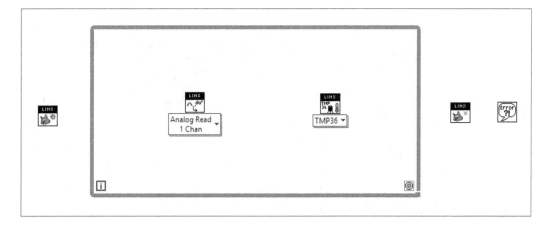

6. Then, we need to connect the different blocks together. To do this, perform these steps:

    1. First, connect the error inputs and outputs together, starting from the left with the initialize element; connect the error output from this element to the input of the next element and so on. After that, do the same with the LINX resource pins, which is the top wire in the following screenshot:

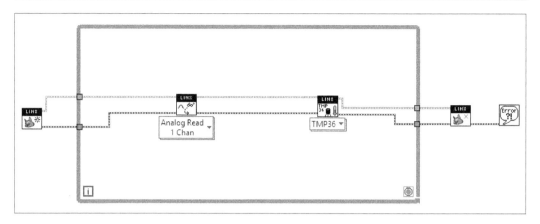

2. The next step is to create the controls to set the parameters of the elements. For the initialize element, you need to create a serial port control to set the correct serial port corresponding to your Arduino board. To do so, simply right-click on the serial port input pin and then go to **Create | Control**.

3. Then, do the same to create controls for the pins of the TMP36 element and the Analog Read element. Also, name these controls so that you know to which element they are connected.

4. Finally, connect the While Loop end condition (the red circle in the bottom-right corner) to the yellow error wire.

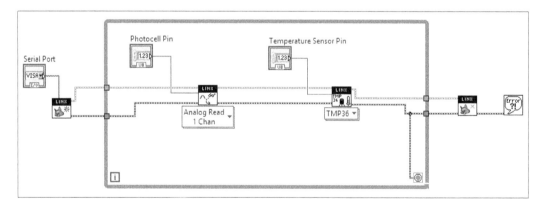

7. We can now create the output of the two central elements in our program. In order to do so, right-click on each element output pin and then go to **Create | Indicator**. Also, name each of the outputs so that we know which measurement is displayed by each indicator.

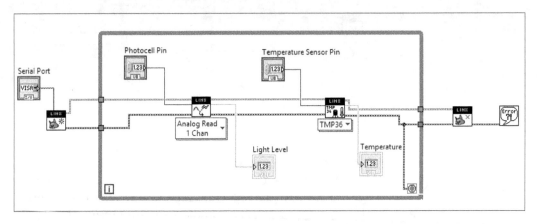

We are done with **Block Diagram** at this point. We can now test the program by going back to **Front Panel**. You will see that you have several elements already on **Front Panel** that correspond to the controls and indicators we created earlier.

You can organize the elements in two categories: the controls on the left-hand side, and the indicators on the right-hand side.

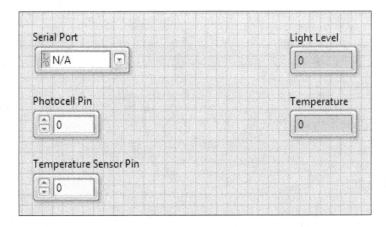

Now, set the controls so that they match our project. To do so follow the steps given:

1. First, select the desired serial port from the **Serial Port** combobox corresponding to your Arduino board; this should be automatically proposed by the LabVIEW software. In my case, it was COM4.

2. Also, set the analog pin to which the photocell is connected (0) and do the same for the TMP36 sensor (1).

3. Then, you can run the program by clicking on the small arrow on the toolbar. You will see that the measurements appear immediately on **Front Panel**:

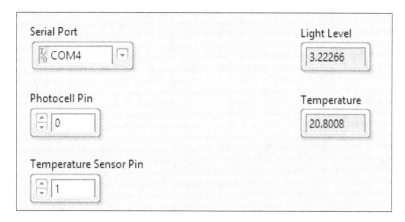

As you can see in the preceding screenshot, the **Temperature** measurement immediately makes sense: **20.8** degrees Celsius. However, we can ask about the value returned by the **Light Level** sensor.

It simply corresponds to the voltage measured on the analog pin on the Arduino board. Indeed, if you play with your hand on top of the sensor, you will see that the value of the light level goes from **0** (complete darkness) to **5** (bright light on the sensor). If the value varies with the ambient light level, it means that the sensor is working correctly.

Upgrading the interface

At this stage, we know that we have our two sensors working and that they were interfaced correctly with the LabVIEW interface. However, we can do better; for now, we simply have a text display of the measurements, which is not elegant to read.

Also, the light-level measurement goes from **0** to **5**, which doesn't mean anything for somebody who will look at the interface for the first time.

Therefore, we will modify the interface slightly. We will add a temperature gauge to display the data coming from the temperature sensor, and we will modify the output of the reading from the photocell to display the measurement from 0 (no light) to 100 percent (maximum brightness).

We first need to place the different display elements. To do this, perform the following steps:

1. Start with **Front Panel**. You can use a temperature gauge for the temperature and a simple slider indicator for **Light Level**. You will find both in the **Indicators** submenu of LabVIEW. After that, simply place them on the right-hand side of the interface and delete the other indicators we used earlier.

2. Also, name the new indicators accordingly so that we can know to which element we have to connect them later.

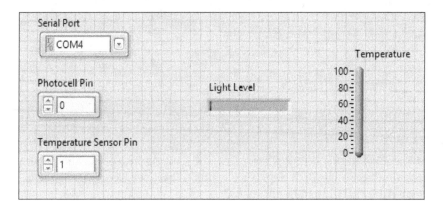

Then, it is time to go back to **Block Diagram** to connect the new elements we just added in **Front Panel**. For the temperature element, it is easy: you can simply connect the temperature gauge to the TMP36 output pin.

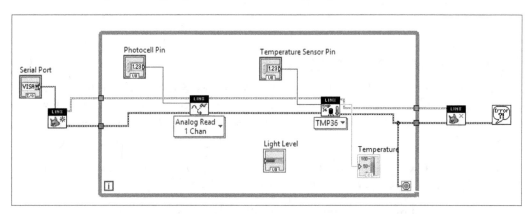

For the light level, we will make slightly more complicated changes. We will divide the measured value beside the **Analog Read** element by 5, thus obtaining an output value between 0 and 1. Then, we will multiply this value by 100, to end up with a value going from 0 to 100 percent of the ambient light level.

To do so perform the following steps:

1. The first step is to place two elements corresponding to the two mathematical operations we want to do: a divide operator and a multiply operator. You can find both of them in the **Functions** panel of LabVIEW. Simply place them close to the **Analog Read** element in your program.

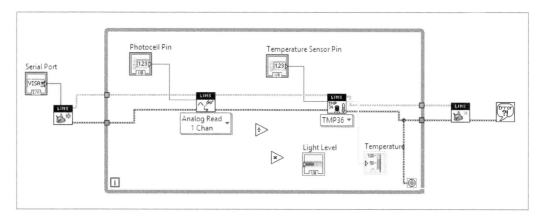

2. After that, right-click on one of the inputs of each operator element, and go to **Create | Constant** to create a constant input for each block. Add a value of 5 for the division block, and add a value of 100 for the multiply block.

3. Finally, connect the output of the **Analog Read** element to the input of the division block, the output of this block to the input of the multiply block, and the output of the multiply block to the input of the **Light Level** indicator.

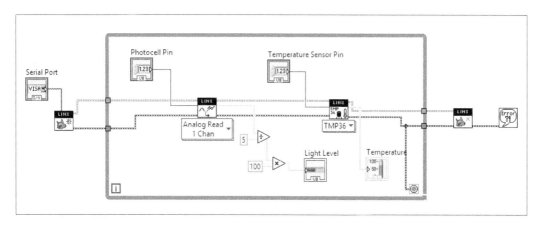

You can now go back to **Front Panel** to see the new interface in action. You can run the program again by clicking on the little arrow on the toolbar.

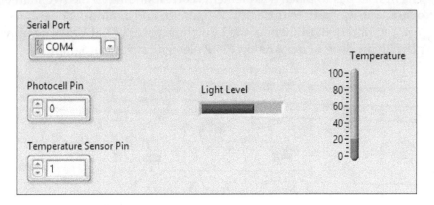

You should immediately see that **Temperature** is now indicated by the gauge on the right and **Light Level** is immediately changing on the slider, depending on how you cover the sensor with your hand.

Summary

Let's summarize what we did in this chapter. We connected a temperature sensor and a light-level sensor to Arduino and built a simple LabVIEW program to read data from these sensors. Then, we built a nice graphical interface to visualize the data coming from these sensors.

There are many ways you can build other projects based on what you learned in this chapter. You can, for example, connect higher temperatures and/or more light-level sensors to the Arduino board and display these measurements in the interface. You can also connect other kinds of sensors that are supported by LabVIEW, for example, other analog sensors. For example, you can add a barometric pressure sensor or a humidity sensor to the project to build an even more complete weather-measurement station.

One other interesting extension of this chapter will be to use the storage and plotting capabilities of LabVIEW to dynamically plot the history of the measured data inside the LabVIEW interface.

In the next chapter, we will build a smart wireless power switch based on Arduino and LabVIEW so that you can control a device remotely and measure its power consumption from LabVIEW.

5
Making an XBee Smart Power Switch

In this chapter, we will build a remotely controlled smart power switch. This switch will be a *do-it-yourself* version of the power switches that you can find in many stores. We will be able to switch a device on and off and also measure the current consumption of the connected device. As the LabVIEW LINX interface for Arduino only supports serial-based communication at the time of writing this book, we will use XBee to communicate wirelessly with the project.

We will first connect the required components to our Arduino board: a relay module, a current sensor, and an XBee shield. Then, we will write a LabVIEW program to control the relay module and measure the current consumption of the connected device. Finally, we will see how to enable XBee communication between your computer and the project.

Hardware and software requirements

Let's first see what we need for this project. Apart from the usual Arduino Uno board, you will need XBee modules, both for Arduino and your computer. Here is the Arduino board with an XBee shield mounted on it, along with one XBee module:

As your computer doesn't come with built-in XBee, you will also need a module on your computer to communicate with the Arduino project via XBee. To do so, I used an USB XBee explorer module from SparkFun, along with an XBee module mounted on it:

To control an electrical device remotely, you will also need a relay module. For this module, I simply used a 5V relay module from Pololu:

For the current sensor, I used a board from ITead Studios, based on the ACS712 chip. Of course, you can use any board based on this chip.

To actually connect an electrical device (I used a 30W desk lamp for this project) to your project, you will need some power cables. You will need two of them: one male power plug that you will use to connect the project to the mains electricity and one female power plug to connect the electrical device to the project. You will also need some electrical wires to make the different connections and some screw terminals to connect the cables together. Here are the two power cables I used for this project:

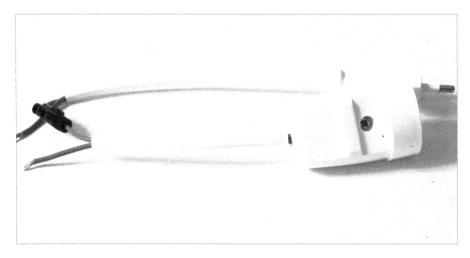

Note that, of course, though you can create the entire project without the power part, which can be done later, the principles remain exactly the same.

This is the list of all the components you will need for this project:

- Arduino Uno (https://www.adafruit.com/products/50)
- Relay module (http://www.pololu.com/product/2480)
- ACS712 current sensor (http://imall.iteadstudio.com/im120710011.html)
- XBee Arduino shield (https://www.sparkfun.com/products/12847)
- XBee module x2 (https://www.sparkfun.com/products/11215)
- XBee explorer module (https://www.sparkfun.com/products/11812)
- Jumper wires (https://www.adafruit.com/products/1957)
- Breadboard (https://www.adafruit.com/products/64)

On the software side, you will need to have LabVIEW and the LINX package installed. If this is not done yet, refer to *Chapter 2, Getting Started with the LabVIEW Interface for Arduino,* to follow all the required steps.

Configuring the hardware

Let's now see how to assemble the different components of the project.

1. Plug the XBee module on the XBee shield and then the XBee shield on the Arduino board. Also, connect the power supply of the Arduino board to the breadboard: connect the 5V pin to the red power rail on the breadboard and the GND pin of the Arduino board to the blue power rail.

2. Then, we will take care of the relay module. The relay module has three pins: VCC, GND, and SIG. First, connect the VCC pin of the relay to the red power rail on the breadboard, the GND pin on the blue power rail, and the SIG pin to the Arduino pin 7.

3. For the current sensor, it is actually quite similar. The current sensor module has three pins as well: VCC, GND, and OUT. First, connect the VCC pin of the current sensor module to the red power rail on the breadboard, the GND pin on the blue power rail, and the OUT pin to the Arduino analog pin A0.

This is what the project looks like without the power cable:

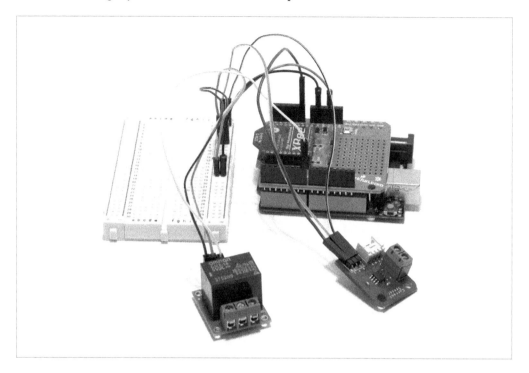

We will now take care of the power part of the project. Using the mains directly is dangerous, and you should ensure that you have all the electrical connections done as described here before plugging the project to the mains. Of course, you can skip this part and come back to it later. Also, never touch the project while it is in use and plugged to the mains; always disconnect the socket from the mains first before changing anything in the project.

Here are the electrical connections that you need to do for this part:

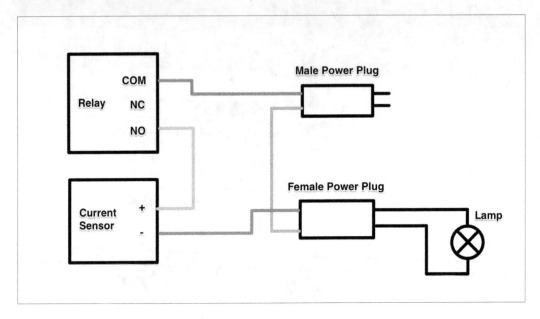

Note that as we are using AC voltages here, you do not need to care about the polarity of the cables. I used a standard 30W desk lamp, but you can use any device of your choice as long as you respect the maximum current/power rating of the relay module you choose. Also, note that the current sensor we are using here is rated at 10A maximum.

Finally, also make sure that you connected the XBee explorer board to your computer via an USB cable.

Controlling the relay

We will now build the first part of our LabVIEW program to control the relay. We basically just want an on/off button to control the relay remotely. For now, connect your Arduino board directly to your computer via USB; we will use XBee later. Also, make sure that the switch on the XBee shield is set to **DLINE**.

As usual, create a new blank VI, place a While Loop, and place one Init block from the LINX interface before the loop and one end block after the While Loop. Also, place a simple error box at the end.

Then, place a **Digital Write** function near the beginning of the loop; we will use this to control the relay. Also, link the LINX resources pins and the error in/out pins together.

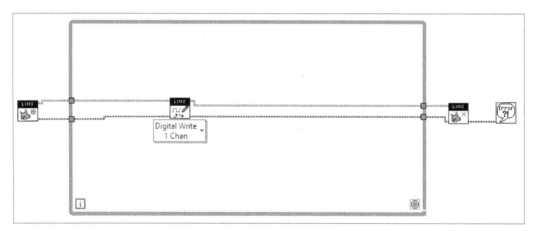

After that, we will create the standard inputs and outputs to make the program work. Start by creating an input for the serial port on the Init block just before the loop. You can create this input by always right-clicking in a blank space of the window and then going to **Create | Control**. Then, create a control for the Arduino pin input of the **Digital Write** block and one control for the input of this block (to control the relay). Finally, don't forget to connect the end condition (the little red circle) of the While Loop to the bottom error wire inside the loop.

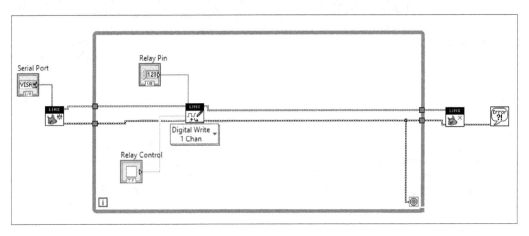

We are now ready to test this very basic program to control the relay. Save the program and go to **Front Panel**. I simply organized the elements as usual: the inputs on the left-hand side and the controls and indicators on the right-hand side.

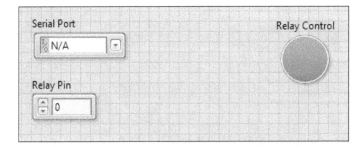

Set the correct **Serial Port** (it should appear in the list) and the correct **Relay Pin** (7), and you are ready to test the program by clicking on the small arrow in the toolbar. Note that it takes 1-2 seconds for the board to be ready, which you can see when both the serial LEDs are activated on the board. After that, you can use the green button. You should hear the relay switch on and off and also see that the device connected to the project is switched on and off accordingly.

If it is not working at this point, there are several things you can check. First, make sure that the LINX interface sketch is still loaded onto your board, and repeat the procedure from *Chapter 2*, *Getting Started with the LabVIEW Interface for Arduino*, if needed. Also, make sure that the relay is connected correctly to the Arduino board, as we saw in the previous section.

Measuring the current

We will now upgrade the LabVIEW program by inserting all the functions required to measure the current consumption of the project. We will first keep things at a basic level; we will just read data from the analog input on which the current sensor is connected and print this data in **Front Panel**.

It starts by inserting an **Analog Read** function (also found in the LINX functions box) just after the **Digital Write** block.

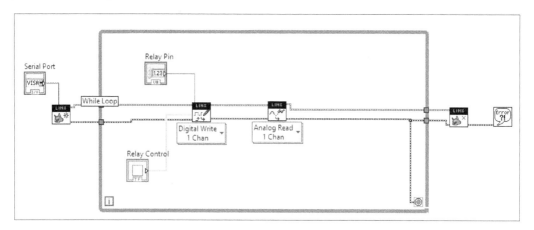

After that, you need to create the relevant input and output for this function; we need to send the pin as an input and get the readout from the Arduino board as an output. As usual, to create an input or output, right-click on the pin and use the **Create** menu. Also. give relevant names to the input and output to know what they correspond to in **Front Panel**.

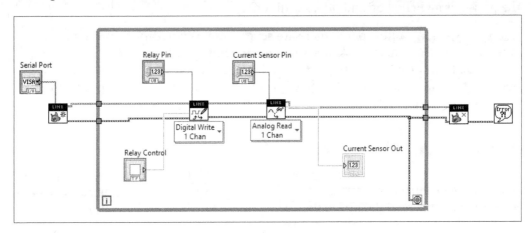

You can now go back to **Front Panel** and organize the elements by adding the **Current Sensor Pin** box on the left-hand side and the **Analog Read** output on the right-hand side. You can now already test the program by clicking on the little arrow in the toolbar. You will see that the readout we added is displaying a value that changes whether the relay is activated or not. This value is directly the output of the **Analog Read** function, so it is between 0 and 5 (as the Arduino is operating between 0 and 5 volts):

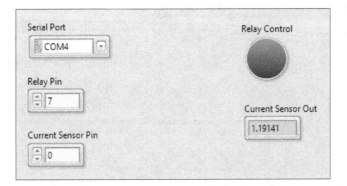

However, this is not very useful yet. For example, if the current flowing through the relay is null, we want 0 to be displayed inside the box. To do so, we will first take a measurement when the program starts and then subtract this value from the live readout coming from the sensor.

This is done by adding another **Analog Read** box before the While Loop and then subtracting the value from this first measurement to the measurement done inside the While Loop.

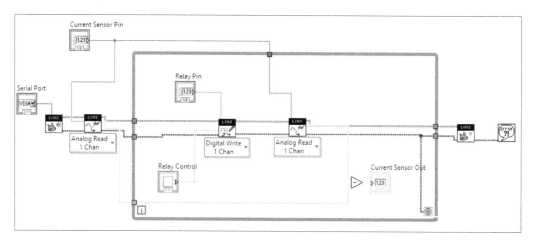

You can now test the program again. This time, the value read by the sensor should oscillate around 0.

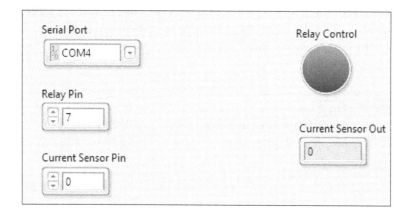

However, this is not perfect yet; we simply have an indication about the current flowing through our switch, but what we would really like to know is how much current is flowing through the switch. We will, therefore, calculate the effective current flowing through our device, using a formula given by the manufacturer:

*Effective_current = analog_measurement / 185 * 1000000 / 1.414*

This formula can be easily translated into LabVIEW by a set of mathematical functions:

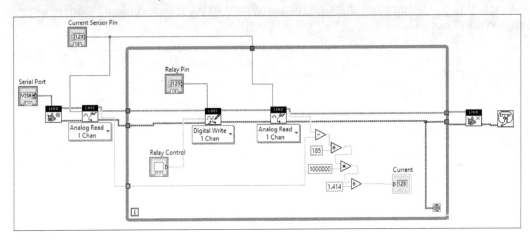

At this point, you can test the program again. However, you will see that the readings are fluctuating a lot. This is due to the analog nature of the sensor that we are using to measure the current.

Therefore, we need to add an additional functionality to our program; we need to make several measurements from the Arduino board, average them, and then use these values to calculate the current consumption of the device. This can be done easily in LabVIEW by using a combination of two elements: one For Loop and second the Mean function.

First, put the **Analog Read** function in the middle of your program inside a For Loop which you can find in the same menu as the While Loop we used earlier. Also, create a constant in front of the little **N** to indicate how many iterations we want to do. I used 100 in this example; this provides a good averaging of readings coming from the board.

Then, search inside the function menu for **Mean** and look for the same **Mean** function as shown in the screenshot:

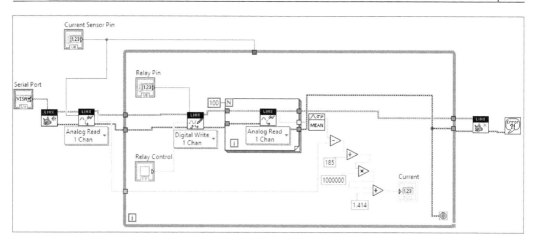

Then, simply place this function after **For Loop** and connect the input of this function to the output of **For Loop** and the output of the **Mean** function (the averaged reading from the Arduino board) to the subtract operator.

You also need to do the same for the first measurement that we are doing before the big While Loop, as shown in the following screenshot:

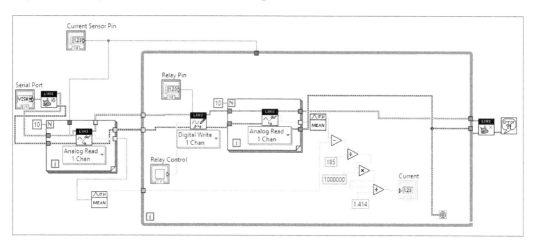

You can now test the project again. This time, you should see that we greatly reduced the fluctuations of the output of the current reading. If this works, congratulations! You just built a smart electrical power switch that you can control from LabVIEW!

Controlling the project via XBee

Finally, we will see how to use the XBee connection on our project to control it remotely from LabVIEW. Luckily, for us, LINX makes it totally transparent to control the project via XBee.

There are just a few things we need to change. First, most of the XBee devices you can buy are set to work at a serial speed of 9600 bauds. Therefore, you will need to change this into the Init function of LINX (the first box we put on the program, just before the While Loop). Double-click on this box, go to **Block Diagram** of this VI, and then add a constant set to 9600 in the INIT DEVICE block to override the default serial speed.

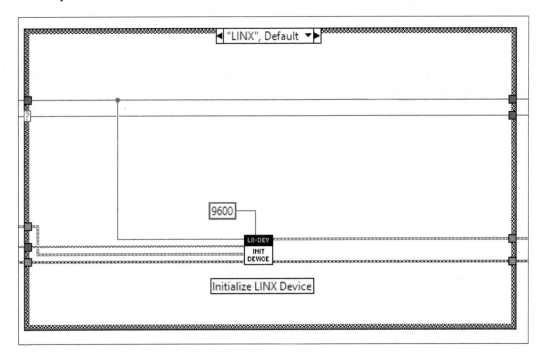

Once this is done, you can go back to **Front Panel** of our Main.vi. Here, you will need to change the Serial port to use the XBee explorer board, and not the serial port from your Arduino board. Simply choose the other serial port and not the one you used earlier.

Before it can work, there is one last step you have to perform: set the little switch on the XBee Arduino shield to UART so that the shield can now send data directly to your Arduino board serial port.

You can now test the sketch again. This time, you will see that the LEDs on both the XBee explorer board and on your XBee shield are blinking; this means that all the communication is now done via XBee.

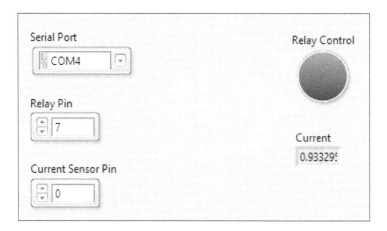

Of course, at this point, you can disconnect the project from your computer, install it wherever you want in your home, and simply power it using the DC power plug on the Arduino board.

Summary

Let's summarize what we did in this chapter. We built a wireless power switch based on Arduino; we controlled this switch using LabVIEW. We connected a relay module, a current sensor, and an XBee shield to an Arduino board to build the hardware part of our project. Then, we built a program from scratch to control the project via LabVIEW.

There are, of course, several ways to improve this project. You can, for example, add more relays and current sensors to the project in order to control two or more devices at the same time. From here, it is really easy to modify the interface to incorporate these news elements.

In the next chapter, we will continue to use XBee, but this time, for a different application: an application to monitor several motion sensors in order to create a simple alarm system.

6
A Wireless Alarm System with LabVIEW

In this chapter, we will use Arduino and LabVIEW to create a simple alarm system that will monitor several motion sensors via XBee. We will connect the motion sensors to an Arduino board, along with an XBee Arduino shield.

Then, we will write a basic LabVIEW program to monitor these motion sensors from LabVIEW. We will start with one sensor only and then extend this to several sensors. Finally, you will see how to use a motion sensor via XBee so that you can place the project wherever you want in your home.

Hardware and software requirements

Let's first see what we need for this project. Apart from the usual Arduino Uno board, you will need XBee modules, both for Arduino and your computer. Here is the Arduino board with an XBee shield mounted on it, along with one XBee module:

As your computer doesn't come with built-in XBee, you will need a module on your computer to communicate with the Arduino project via XBee. To do so, I used an USB XBee explorer module from SparkFun, along with an XBee module mounted on it:

Note that for this chapter, you can also use a Bluetooth module to interface your Arduino board with your computer, without any major change in the code.

Finally, you will also need one or many motion sensors. I used a simple PIR motion sensor for this project; you can find this sensor on many resellers' websites. The most important point here is that the sensor should be compatible with the Arduino Uno voltage levels, that is, the sensor should have a maximum voltage output of 5V.

This is the list of all the components you will need for this project:

- Arduino Uno (`https://www.adafruit.com/products/50`)
- Motion sensor (`http://www.adafruit.com/products/189`)
- XBee Arduino shield (`https://www.sparkfun.com/products/12847`)
- XBee module x2 (`https://www.sparkfun.com/products/11215`)
- XBee explorer module (`https://www.sparkfun.com/products/11812`)
- Jumper wires (`https://www.adafruit.com/products/1957`)

On the software side, you will need to have LabVIEW and the LINX package installed. If this is not done yet, refer to *Chapter 2*, *Getting Started with the LabVIEW Interface for Arduino*, to follow all the required steps.

Hardware configuration

We will now assemble the different components of the project by performing the following steps:

1. Plug the XBee module on the XBee shield and then the XBee shield on the Arduino board.
2. Then, connect a motion sensor. Of course, you will need to repeat the operation if you want to connect several motion sensors to the board.
3. A motion sensor has three pins: VCC, GND, and OUT (or SIG). First, connect the VCC pin to the Arduino 5V and the GND pin to the Arduino GND pin.
4. Connect the OUT pin of the motion sensor to pin number 8 on the Arduino board.

This is how the fully assembled project looks, with one motion sensor connected:

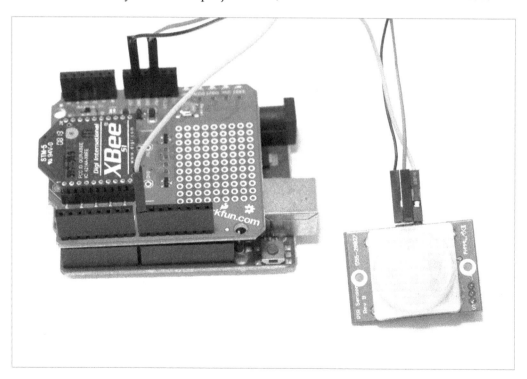

Finally, also make sure that you connected the XBee explorer board to your computer via a USB cable.

Interfacing one motion sensor

Now that the hardware part of the project is configured, we will write a basic LabVIEW program to handle a one motion sensor and display its state in the LabVIEW interface. The steps are as follows:

1. Start by creating a new blank VI and add the required components for any LINX program: one LINX Init function, one LINX stop function, and one simple error box at the end. Also, add a While Loop in the middle of the VI; here, we will add the functions to read the data that comes from the motion sensor.

2. The motion sensor we are using for this project simply returns a logic level of 0 if no motion is detected; otherwise, it returns a logic level of 1. To read data from the motion sensor, we need a **Digital Read** function. As usual, you can find this function inside the LINX submenu when placing a function from LabVIEW. Place this function in the middle of the While Loop:

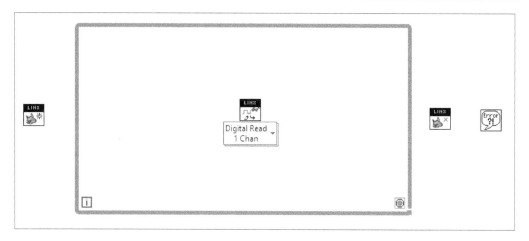

3. Connect the LINX resource pins together (the top wires of the screenshot), and also connect the error inputs/outputs together (the bottom wires). Also, connect the While Loop end condition (the little circle in the bottom-right corner) directly to the wire of the error link, which is the wire at the bottom of this screenshot:

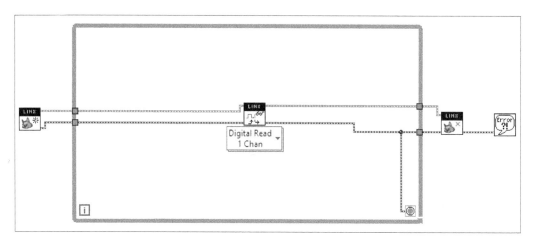

We can now add some inputs and outputs to our program. We need two inputs: one to set the serial port of the Arduino board and one to set the pin on which the motion sensor is connected.

To do so, right-click on the desired pin and then go to **Create | Control** to create a new input. Also, give relevant names to these inputs so that we can easily identify them from the front panel.

We also need to create one output for the motion sensor. To do so, right-click on the output of the **Digital Read** block and go to **Create | Indicator**. Also, give a name to this indicator to show that it is the output of the motion sensor.

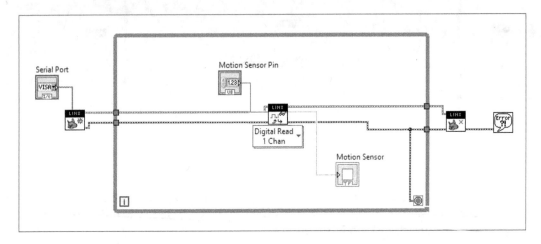

4. We are now ready to test the program we just wrote. Go to the **Front Panel** and organize the elements as usual: controls on the left and indicators on the right, just as shown in the following screenshot:

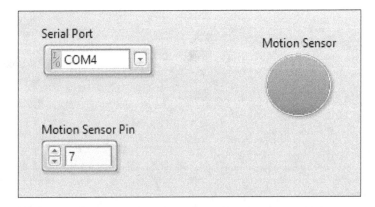

Of course, you need to set the correct **Serial Port** by selecting it from the list and also set the **Motion Sensor Pin** to the pin we used earlier, that is, **7**.

You can now start the program by clicking on the little arrow on the toolbar. Give it 1-2 seconds to initialize, and then, pass your hand in front of the motion sensor. You will see that the indicator turns green immediately. After a while, if no motion is detected anymore, it will return to normal.

Connecting more motion sensors

Now that we have one motion sensor working, we will interface more of them with our Arduino board and LabVIEW.

First, make sure that you have more than one sensor connected to your Arduino board. If necessary, repeat the hardware configuration procedure again. For example, in the following screenshot, I used two motion sensors: one connected to pin number 7, as we did earlier, and one connected to pin number 8.

We will now modify the LabVIEW program to accommodate two sensors. Of course, if you have more of them connected, you could simply connect the sensor as you did before for each additional sensor you have.

To add another sensor, you need to repeat the procedure we did earlier. First, create a new **Digital Read** function and connect the LINX resource wires (the top wires) and the error input/output wires to the new box so that it looks like this screenshot:

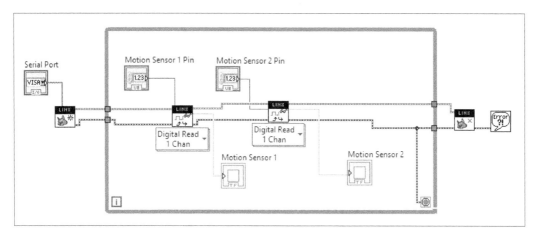

Also, add a new input for this new box (to set the pin of this new motion sensor) and one indicator as an output. Make sure that you give relevant names to these new inputs and outputs as well, as we now have two motion sensors in the **Front Panel**.

You can now go back to the **Front Panel** to see the result. As you can see, we now have one additional input and one additional output; this is the state of the new motion sensor:

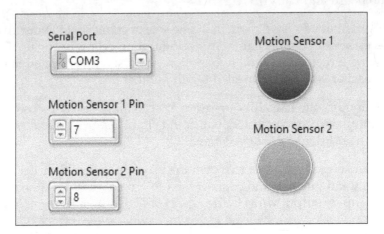

As usual, I recommend that you organize the **Front Panel** with all the inputs on one side and all the indicators on the other side.

Now, we can test the program again. Make sure that the serial port is still correctly set and that you correctly set the motion sensors' pins. Then, click on the little arrow on the toolbar to run the program.

You will see that now, both motion sensors are reacting independently, depending on which motion sensor is in the detection mode.

Making the project wireless with XBee

In this final section of the chapter, we will use the XBee shield that we connected to the project earlier in order to control a wireless alarm system from LabVIEW.

The first step to use XBee, instead of the USB connection, is to set the little switch on the XBee Arduino shield to the **UART** position. Earlier, it was on **DLINE**, just s shown in the image:

Then, we need to specify again that we are using a serial speed of 9600 bauds for the XBee module, just as we did in the previous chapter. Indeed, we need to do this as the XBee modules are configured to use a speed of 9600 bauds by default. To know more about UART communications, you can visit the following resource http://en.wikipedia.org/wiki/Universal_asynchronous_receiver/transmitter.

To change this, simply double-click on the **Init** function that we placed just before the While Loop in the diagram. This will open the VI.

Once this is done, go to the **Block Diagram** of this VI and look for a function called **INIT DEVICE**. Once you find it, create a constant for the pin called Override Serial Speed and enter a value of 9600, just as shown in the following screenshot:

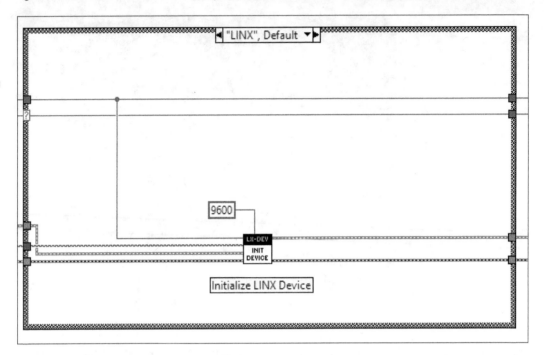

You can now save this VI and close it. You can also return to the VI that contains the program for the two motion sensors. You now simply need to select the serial port that corresponds to your XBee USB explorer board. To know which serial port you need to select, simply take the one you did not use earlier when working with the project via the USB port.

You can now start the program again by clicking on the little arrow on the top toolbar. You will notice that the serial LEDs on your XBee shield and on the explorer board are now constantly blinking; this means that communication was established via XBee.

At this point, pass your hand again in front of one sensor; you will see that the corresponding indicator immediately changes on the **Front Panel**.

Summary

We created a simple alarm system based on Arduino and controlled it via LabVIEW. We interfaced motion sensors to our Arduino board and monitored the state of these sensors via LabVIEW. We also used XBee modules to monitor our alarm system wirelessly.

There are, of course, several ways to go further with this project. You can add even more motion sensors to the projects. However, at the time of writing this book, it was not possible to include several XBee modules and have several distant Arduino projects communicating with a central LabVIEW interface. However, we can also make the project more visual, for example, by connecting an LED to one pin of the Arduino board and making this LED blink every time motion is detected on one motion sensor.

In the next and final chapter of the book, you will use everything you learned so far to build an Arduino-based mobile robot that you will control using LabVIEW.

7
A Remotely Controlled Mobile Robot

In the final chapter of this book, you will take everything you have learned so far about Arduino and LabVIEW to build a remotely controlled mobile robot. We will build a robot using a kit that is already made, add some Arduino boards and a wireless module, and finally, control everything from LabVIEW.

Hardware and software requirements

Let's take a look at the components required for this robot. The first thing that we will need is the robot chassis itself. You have several options for this part, but I recommend the DFRobot MiniQ 2 wheels chassis. It is a tiny robot with two motors and two wheels and is compatible with most Arduino boards. I chose this robot for its low price point and because it is really easy to use with Arduino. However, you can choose any robot chassis that has two wheels and that is compatible with Arduino boards.

You will then require several Arduino boards to interface with the robot. The first board is the usual Arduino Uno board, which will serve as the brain of the robot. You will also need a DFRobot motor shield to control the two wheels of the robot from Arduino. Finally, you will need an Arduino XBee shield in order to mount an XBee Series 1 module. This will be used to control the robot remotely.

After that, you will need an URM37 Ultrasonic Sensor to measure the distance in front of the robot. To get more information on this sensor, you can visit the following link to the manual `http://www.dfrobot.com/wiki/index.php?title=URM37_V3.2_Ultrasonic_Sensor_(SKU:SEN0001)`

You will also need a LiPo battery to power up the robot. For this purpose, we will utilize a 7.2V LiPo battery.

We will also use an XBee USB controller with another XBee Series 1 module to control the robot from the computer.

The following list shows all the components required for this project:

- Arduino Uno (https://www.adafruit.com/products/50)
- DFRobot MiniQ robot chassis (http://www.dfrobot.com/index.php?route=product/product&search=miniq&description=true&product_id=367)
- DFRobot motor shield (http://www.dfrobot.com/index.php?route=product/product&product_id=59&search=motor+shield&description=true)
- URM37 ultrasonic sensor (http://www.dfrobot.com/index.php?route=product/product&product_id=53&search=ultrasonic&description=true)
- 7.4V LiPo battery (http://www.dfrobot.com/index.php?route=product/product&product_id=489&search=battery&description=true)
- Arduino XBee shield (https://www.sparkfun.com/products/12847)
- XBee module x2 (https://www.sparkfun.com/products/11215)
- XBee explorer module (https://www.sparkfun.com/products/11812)
- Jumper wires (https://www.adafruit.com/products/1957)

On the software side, you will need to have LabVIEW and the LINX package installed. If this is not done yet, refer to *Chapter 2, Getting Started with the LabVIEW Interface for Arduino*, to follow all the required steps.

Hardware configuration

We will now assemble the different components of the project by performing the following steps:

1. Assemble the robot chassis using the instructions given by the manufacturer of the chassis. Then, assemble the metal spacers to mount the Arduino Uno board and other boards later. Finally, mount the ultrasonic sensor in front of the robot, as shown in the image:

2. The next step is to assemble the Arduino Uno board on the robot. Mount the Arduino Uno board on top of the spacers. Then, screw it to the spacers with at least two screws for a better hold so that it is held firmly in place.

3. We can now mount the motor shield. Put it on top of the Arduino Uno board that we assembled earlier. Then, connect it to the motors via the screw terminals. Make sure that you are using the same polarity for each motor. For example, if you have connected the red wire from motor 2 to the M2+ header, connect the red wire from motor 1 to the M1+ header.

4. We will now mount the ArduinoXBee shield on the robot and the XBee module on top of the shield:

5. Let's now interface the ultrasonic sensor with the robot. There are three pins we need to connect for this sensor: VCC, GND, and the pulse output from the sensor. If we look at the sensor from the pins' side and starting from the left-hand side, the first pin is the VCC pin, the second pin is the GND pin, and the fourth pin corresponds to the pulse output. Connect VCC to the Arduino 5V pin, GND to Arduino GND, and finally, the pulse output of the sensor to Arduino pin 3.

6. Before using the robot with LabVIEW, connect the battery to the DC jack input of the Arduino Uno board.

7. Finally, plug the XBee USB explorer module with an XBee Series 1 module on it to your computer.

Moving the robot around

We will now build a simple LabVIEW sketch to control the wheels of the robot. You will be able to set the speed of the robot and change the direction of each wheel.

1. It starts as usual with an empty While Loop where we will add all our blocks to control the robot. We can also place the usual Init and stop blocks, and a simple error box at the end:

2. We will place four functions inside the main While Loop, two per wheel of the robot. Indeed, for each wheel, we need two functions per wheel/motor: one **Digital Write** function to set the direction of the motor and one **Set Duty Cycle** function to set the speed the motor. Place these blocks inside the loop, and connect the usual LINX resource wires and error links so that it looks like the following screenshot:

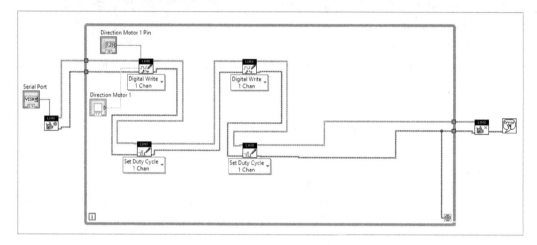

3. In the preceding screenshot, we already created some control blocks for the first **Digital Write** function, and we connected the end condition of the While Loop to the error wire. Also, we already created a **Serial Port** control for the corresponding input of the Init block.

4. You can now create more control for each input of the functions we just added in the While Loop by right-clicking on each input. Note that for each function, you need to create two controls: one for the value of the input itself and one for the pin. Also, make sure that you give relevant names to each control you place at this step, as it will be useful when placing the different elements on **Front Panel**.

We can now go back to **Front Panel**. As usual, I have added all the controls of the pins on the left-hand side and the control of the value of the functions on the right-hand side.

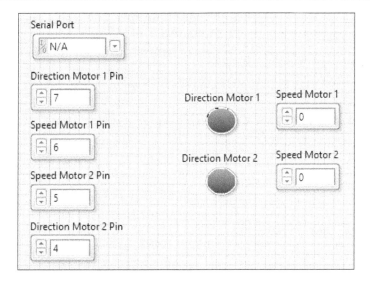

It is now time to try our robot interface for the first time. First, choose the correct serial port from the list and make sure that the little switch is set to **DLINE** on your ArduinoXBee shield.

Then, you have to set the correct pins for the motor shield; you can find out about these pins in the DFRobot motor shield documentation at this link http://www.dfrobot.com/wiki/index.php?title=Arduino_Motor_Shield_(L293)_(SKU:_DRI0001)

We can now test the robot. Make sure that the battery is connected to the Arduino Uno board, and click on the little arrow on the toolbar to start the program. Also, make sure that the robot has the wheels in the air, because it is still attached via the USB cable to your computer.

Then, you can try to enter values in the speed controls (between 0 and 255), and you should see that the wheel immediately starts to rotate. You can also change the direction of a given wheel by clicking on one of the green buttons shown here:

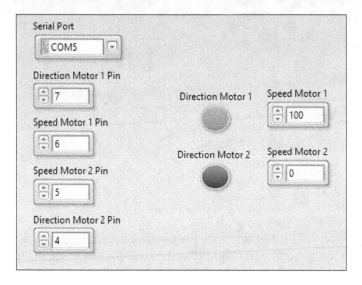

We will now improve this interface a little bit. Indeed, it is not convenient to change the speed of the robot by writing down the speed. Instead, we will use sliders to control the robot with just our mouse.

Still on the **Front Panel**, remove the two text inputs for speed and add two sliders with pointers instead. Name these new controls, and open their **Properties** panels to change the maximum value of the sliders to 255.

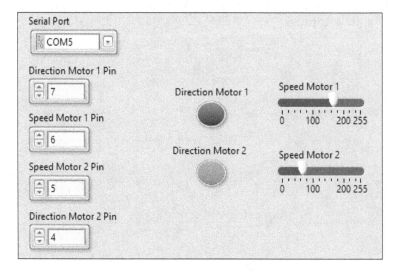

Of course, you will then have to go back to **Block Diagram** to connect the new controls to the **Set Duty Cycle** boxes.

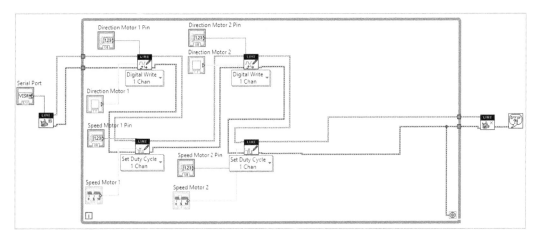

You can now go back to the Front Panel and test the new controls; you will now be able to control the speed of the robot just by dragging the cursors of the sliders.

Measuring the front distance

We will now upgrade the program we just wrote to include the front ultrasonic sensor. We will simply display the value measured by the sensor in the Front Panel of our program.

The first step is to make the appropriate modifications to the **Block Diagram**. To measure the distance in front of the sensor, we will use a pulseIn block that you can find in the LINX toolbox. Basically, the sensor will return a pulse whose length is proportional to the distance in front of the sensor.

Place the `pulseIn()` function in the remaining space inside the While Loop, and then, reconnect the LINX resource wire and the error wire so that they integrate with this new function. Also, create controls for the pin of the `pulseIn()` function, and set the function's upper-left pin to **Active Low** by creating a new control.

For the output, we first need to divide the output of the box by 50 to get a reading in centimeters. Use a simple divide function for this. After this divide function, create a simple text output to display the measured value in centimeters. The following screenshot summarizes all the changes made at this point:

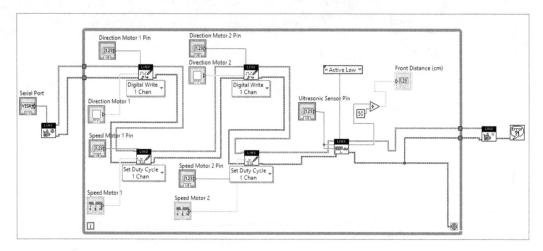

To help you out, this screenshot is a zoom-in view of the new functions we added to integrate the ultrasonic distance sensor:

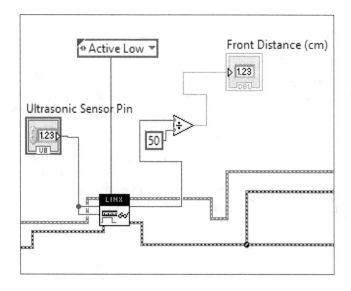

We can now go back to the **Front Panel**. The first step here is to arrange the different controls again and add the output text from the sensor on the far right part of the panel. Also, set the pin of the `pulseIn()` function to the correct value (3).

We can now test the program again. Simply run it and watch the distance measured by the sensor; it will be updated in real time, as you move your hand in front of it.

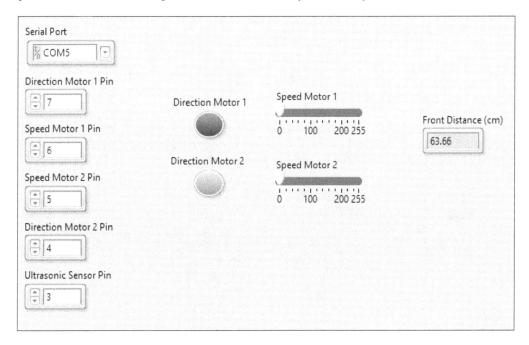

Controlling the robot wirelessly

We now have a functional robot, as we interfaced all the motors and sensors with LabVIEW. However, we are still stuck with the USB cable to send data to the robot, so it makes it impractical to move the robot around. In this last part of the chapter, we will quickly see how to use the XBee module that we installed on top of the robot to control it remotely.

You can already disconnect the USB cable from the robot and make sure that the battery is still connected. Also, make sure that the XBee USB explorer board is connected to your computer.

Then, go inside the Init box of the program and open its **Block Diagram**. Make sure that the serial speed is set to 9600 bauds, just as shown in the following screenshot:

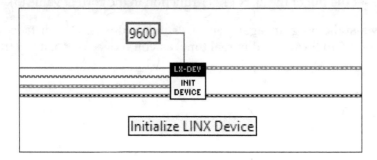

Now, go back to the **Front Panel** of our main program and change the serial port so that it matches the serial port of the XBee explorer board. Finally, also make sure that the switch on the ArduinoXBee shield is set to **UART**.

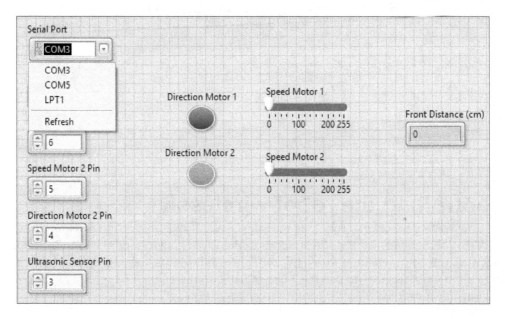

You can now run the program again. You will see that the LEDs on the ArduinoXBee shield and explorer board are blinking continuously; this means that communication has been established between both. You can now control your mobile robot without any wires!

Summary

You learned how to build an Arduino-powered robot, which can be controlled remotely via XBee. We also interfaced this robot with LabVIEW, first to move it around and then, to measure the distance in front of the robot via an ultrasonic sensor.

There are, of course, several ways to improve this project. The first one will be to add more sophisticated controls for the robot. For example, you can use the basic command blocks we defined for the robot to create controls that make the robot move forward, backward, turn left, and so on. You can also create automated commands inside the program, for example, to make the robot go backward when it detects an obstacle in front of the sensor.

I hope that this book gave you a good overview of what is possible using LabVIEW along with Arduino. The possibilities are endless, and you can create really exciting projects using LabVIEW to control your Arduino projects, and all this without writing one line of code! At the time of writing this book, the LINX toolbox is still limited to the basics, but as it improves in the future; it will allow us to build even more amazing projects using the combination of LabVIEW and Arduino.

Index

hardware, configuring 48-50
project, controlling 58, 59
relay, controlling 50-53

I

integrated development
    environment (IDE) 5

J

Jumper wires
  URL 24

L

L293D
  URL 24
  URL, for configuration 25
LabVIEW
  Arduino, features 6
  DC motor, controlling 23
  downloading 7, 8
  features 6
  installation, testing 17-21
  setting up 14-16
  skill requisites 6
  software, reference link 26
  URL 7, 11
  using, with Arduino 6
LabVIEW Interface for Arduino (LIFA)
  hardware requisites 11-13
  software requisites 11-14
  URL 12
LINX
  installation, testing 17-21
  setting up 14-16

M

Motion sensor
  URL 62

N

NI-VISA package
  URL 19

P

photocell
  URL 36
project
  controlling, via XBee 58, 59
Pulse Width Modulation (PWM)
  reference link 27

R

Relay module
  URL 48
remotely controlled mobile robot
  hardware and software requisites 73

T

TMP36
  URL 36

U

UART communications
  URL 69
URM37 Ultrasonic sensor
  URL 73, 74

V

virtual instruments (VIs) 6
Visual Package Manager (VIPM)
  about 5
  URL 12

W

weather station
building 35
hardware and software requisites 35, 36
wireless alarm system, LabVIEW
hardware and software requisites 61, 62

X

XBee
project, controlling via 58, 59
XBee Arduino shield
URL 48, 62

XBee explorer module
URL 48, 62, 74
XBee module x2
URL 48
XBee Smart Power Switch
hardware and software requisites 46-48

Thank you for buying
Programming Arduino
with LabVIEW

About Packt Publishing

Packt, pronounced 'packed', published its first book, *Mastering phpMyAdmin for Effective MySQL Management*, in April 2004, and subsequently continued to specialize in publishing highly focused books on specific technologies and solutions.

Our books and publications share the experiences of your fellow IT professionals in adapting and customizing today's systems, applications, and frameworks. Our solution-based books give you the knowledge and power to customize the software and technologies you're using to get the job done. Packt books are more specific and less general than the IT books you have seen in the past. Our unique business model allows us to bring you more focused information, giving you more of what you need to know, and less of what you don't.

Packt is a modern yet unique publishing company that focuses on producing quality, cutting-edge books for communities of developers, administrators, and newbies alike. For more information, please visit our website at www.packtpub.com.

Writing for Packt

We welcome all inquiries from people who are interested in authoring. Book proposals should be sent to author@packtpub.com. If your book idea is still at an early stage and you would like to discuss it first before writing a formal book proposal, then please contact us; one of our commissioning editors will get in touch with you.

We're not just looking for published authors; if you have strong technical skills but no writing experience, our experienced editors can help you develop a writing career, or simply get some additional reward for your expertise.

Arduino Home Automation Projects

ISBN: 978-1-78398-606-4   Paperback: 132 pages

Automate your home using the powerful
Arduino platform

1. Interface home automation components
   with Arduino.

2. Automate your projects to communicate
   wirelessly using XBee, Bluetooth and WiFi.

3. Build seven exciting, instruction-based home
   automation projects with Arduino in no time.

Arduino Android Blueprints

ISBN: 978-1-78439-038-9   Paperback: 250 pages

Get the best out of Arduino by interfacing it with
Android to create engaging interactive projects

1. Learn how to interface with and control
   Arduino using Android devices.

2. Discover how you can utilize the combined
   power of Android and Arduino for your
   own projects.

3. Practical, step-by-step examples to help you
   unleash the power of Arduino with Android.

Please check **www.PacktPub.com** for information on our titles

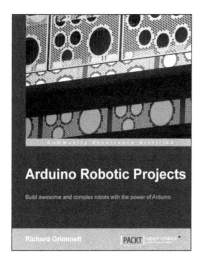

Arduino Robotic Projects

ISBN: 978-1-78398-982-9   Paperback: 240 pages

Build awesome and complex robots with the power of Arduino

1. Develop a series of exciting robots that can sail, go under water, and fly.

2. Simple, easy-to-understand instructions to program Arduino.

3. Effectively control the movements of all types of motors using Arduino.

C Programming for Arduino

ISBN: 978-1-84951-758-4   Paperback: 512 pages

Learn how to program and use Arduino boards with a series of engaging examples, illustrating each core concept

1. Use Arduino boards in your own electronic hardware and software projects.

2. Sense the world by using several sensory components with your Arduino boards.

3. Create tangible and reactive interfaces with your computer.

4. Discover a world of creative wiring and coding fun!

Please check **www.PacktPub.com** for information on our titles